iPhone 17 User Guide for Seniors and Beginners

The Complete Step-by-Step Manual to Master Your iPhone 17, Explore iOS 18 Made Simple, and Unlock Hidden Features with Ease

Georgette Howard

1

Table of Contents

Disclaimer

This book is an independent publication and is not authorized, sponsored, or endorsed by Apple Inc. "iPhone" and "iOS" are trademarks of Apple Inc. Any references to specific features, apps, or updates are for educational purposes only. Every effort has been made to ensure accuracy, but software changes and updates may alter the steps shown.

Preface

The iPhone 17 is not just a phone—it is the most personal, powerful, and versatile device Apple has ever released. Yet for many new users, especially seniors and beginners, this cutting-edge technology can feel overwhelming at first. That is exactly why this guide exists: to make learning the iPhone 17 simple, clear, and enjoyable.

When Apple launched the iPhone 17, iPhone 17 Pro, and iPhone 17 Pro Max, millions of people searched for a manual that explained things in plain English. Apple provides a reference guide, but what most users need is a step-by-step user guide that speaks their language, shows pictorial explanations, and offers practical shortcuts for everyday use.

This book is written with that purpose in mind. Whether you are switching from Android to iPhone 17, picking up your first smartphone, or upgrading from an older model, you'll find everything you need here:

- How to set up your iPhone 17 with confidence
- Clear illustrations showing buttons, gestures, and menus
- Easy instructions for making calls, sending messages, FaceTime, and email
- Tips for taking professional-quality photos and 4K videos
- Guides to downloading apps, browsing the internet, and staying safe online
- Accessibility features designed especially for seniors, beginners, and visually impaired users
- Bonus sections including iOS 18 tips & tricks, hidden features, troubleshooting, and must-have apps

What makes this guide different is its unique focus on seniors and first-time users. Every chapter is written with larger, easy-to-read explanations, plain words instead of technical jargon, and KDP-safe pictorial illustrations with captions that show you exactly where to tap, swipe, or press.

Along the way, you'll also discover hidden iPhone 17 features like Back Tap shortcuts, Action Button customization, Live Text, Translate, Siri commands, and improved battery management. These are the kinds of tips that turn a new phone from something intimidating into something empowering.

By the end of this guide, you won't just know how to use your iPhone 17—you'll feel confident, independent, and ready to explore the digital world at your own pace.

If this is your first iPhone, welcome to the family. If you've used earlier models, get ready to unlock the best that the iPhone 17 series and iOS 18 have to offer. This book is your personal companion every step of the way.

Introduction

Congratulations on your new iPhone 17! Whether you've chosen the iPhone 17, iPhone 17 Pro, or iPhone 17 Pro Max, you now own one of the most powerful and user-friendly smartphones ever created. For many, though, the excitement of owning a new iPhone quickly gives way to confusion. Where do you start? How do you set it up? What do all these icons and gestures mean?

You are not alone. Each year, millions of new iPhone users—especially seniors and beginners—feel the same way. Apple's devices are packed with incredible features, but without the right guidance, they can seem overwhelming. That's why this book was created: to serve as your step-by-step iPhone 17 manual, written in plain language, with illustrations, tips, and real-world examples that make everything simple.

Inside these pages, you'll learn how to:

- Set up your iPhone 17 from the very first screen with ease

- Understand the difference between iOS 18 features, gestures, and the Action Button

- Make calls, send texts, use FaceTime, and set up your email accounts

- Take amazing photos and videos using Portrait mode, Night mode, and Cinematic 4K recording

- Stay connected with apps like WhatsApp, Zoom, Facebook, and Kindle

- Keep your phone safe with Face ID, passcodes, two-factor authentication, and scam protection tips for seniors

- Customize your iPhone with wallpapers, widgets, and accessibility settings that fit your needs

- Use hidden features like Back Tap, Live Text, Translate, and Siri to save time every day

- Solve common problems with our simple troubleshooting guide

This book is more than a manual—it's like having a patient friend sitting beside you, showing you exactly where to tap, swipe, and press until it all feels natural. Each chapter is carefully designed to build your confidence, with **pictorial explanations and captions** that ensure you never feel lost.

For seniors and beginners, this guide also includes:

- Larger, easy-to-read explanations
- Accessibility tips for bigger text, magnifier, and voice dictation
- A **Senior Quick Reference Guide** with essential icons and gestures at a glance
- A simple glossary of iPhone terms explained in plain English

The iPhone 17 is meant to be more than a device—it's a tool to help you stay connected with loved ones, capture priceless memories, learn, shop, and even manage your health. By the end of this book, you won't just know how to use your iPhone 17—you'll feel confident, empowered, and ready to enjoy everything it has to offer.

Welcome to your new journey with the iPhone 17 and iOS 18. Let's begin.

Chapter 1

Meet Your iPhone 17

Your new iPhone 17 is more than just a phone—it's a camera, a diary, a personal assistant, and a gateway to staying connected with friends and family. In this chapter, you'll discover what's new in the iPhone 17, how to choose the right model for your needs, and what to expect when you open the box for the very first time.

What's New in iPhone 17

Apple designed the iPhone 17 to be faster, smarter, and more helpful than ever before. Here are the standout features explained in simple, everyday terms:

- **Sharper Camera System** – With improved night mode and higher-quality zoom, your photos come out brighter and clearer, even in low light.

- **Action Button** – A new customizable button on the side lets you set quick shortcuts—like opening the flashlight, camera, or even recording a voice note.

- **Longer Battery Life** – Lasts up to 2 hours more than the iPhone 16, keeping you powered through the day.

- **iOS 18 Software** – Apple's newest system adds smarter Siri suggestions, lock screen widgets, and new ways to customize your Home Screen.

- **Brighter Display** – Easier to read outdoors, even in strong sunlight.

Tip: *Don't worry if these features sound technical. This book will explain how to use each one step by step.*

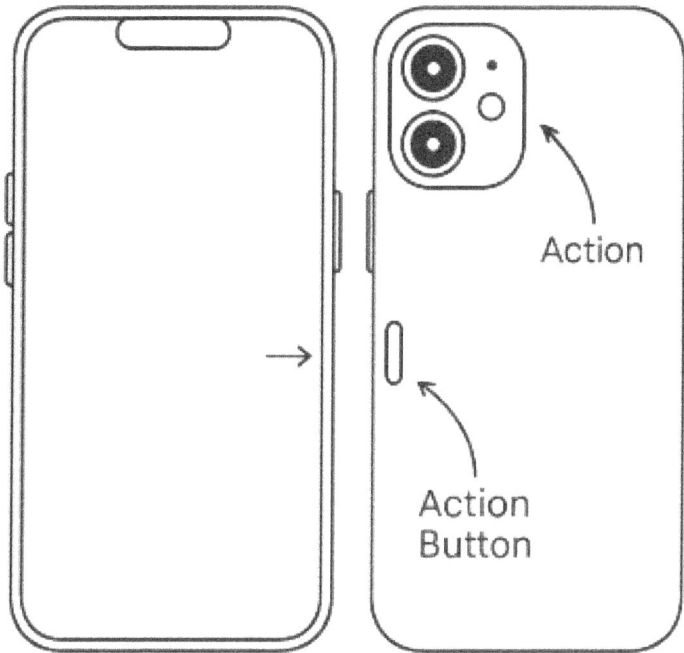

The iPhone 17 comes with new features like Action Button and an upgraded camera system.

Choosing the Right Model

The iPhone 17 comes in four versions. The differences are mainly about size, camera strength, and price.

1. **iPhone 17 (Standard)** – The regular-sized model. Perfect if you want something lightweight, pocket-friendly, and reliable for everyday use.

2. **iPhone 17 Plus** – Slightly larger screen and longer battery life. Great for those who watch videos, read books, or prefer a bigger display.

3. **iPhone 17 Pro** – Includes advanced cameras (telephoto zoom, pro photography features) and a tougher build. Perfect for photography lovers.

4. **iPhone 17 Pro Max** – The largest screen, best cameras, and the longest battery life. Designed for people who want the top-of-the-line iPhone experience.

Tip for Seniors: *If you have difficulty with small text, you may prefer the Plus or Pro Max for the larger display.*

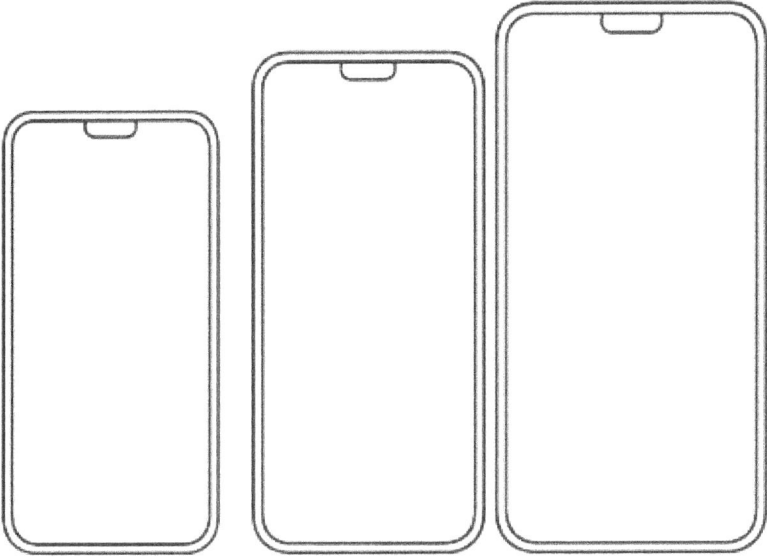

iPhone 17 comes in four versions.
The larger the model, the
bigger the screen and batterylife.

Unboxing: What's in the Box

When you open your iPhone 17 package, here's what you'll find inside:

- **The iPhone 17 device** – Carefully wrapped for protection.

- **USB-C Charging Cable** – Apple now uses a universal USB-C connector, making it easier to charge and connect to other devices.

- **Documentation** – Small booklets with warranty and safety information.

- **SIM Ejector Tool (in some regions)** – Lets you insert or remove the SIM card tray.

Important Note: Apple no longer includes a wall charging adapter or wired earphones in the box. You may need to purchase these separately if you don't already own them.

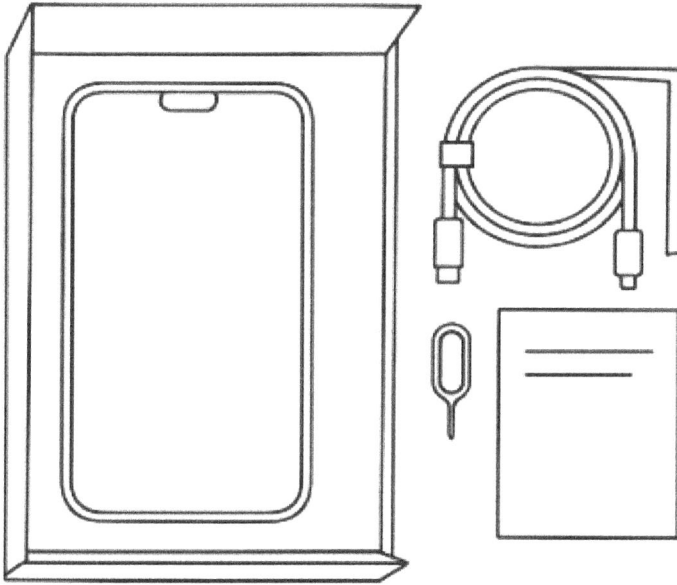

Your iPhone 17 box includes the device,
charging cable, and small documents
The power adapter is sold
separately.

By the end of this chapter, you now know the basics of
what makes the iPhone 17 special, which model might fit
you best, and what to expect when you open the box. In
the next chapter, we'll walk step by step through setting
up your iPhone 17 for the very first time.

Side button — Cameras

Volume rocker — Flash

Lightning port — LiDAR scanner

Lightning port

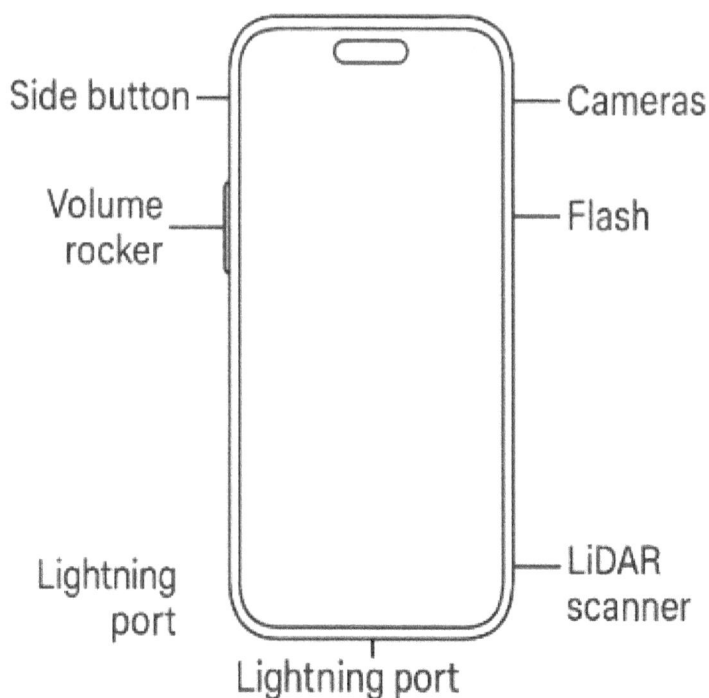

Know your phone at a glance: the side button powers on, the volume rocker adjusts sound, the action button customizes shortcuts.

Chapter 2

Getting Started for the First Time

You've unboxed your brand-new iPhone 17. Now let's bring it to life. This chapter will walk you through powering it on, setting up the basics, and making sure accessibility features are ready so your phone is truly comfortable for you.

Powering On Your iPhone 17

1. Hold down the **Side Button** (on the right edge of your phone).
2. The Apple logo will appear on the screen.
3. Wait a few seconds until the **"Hello" welcome screen** greets you.

Tip: *If your phone doesn't turn on, charge it for at least*

10–15 minutes using the USB-C cable before trying again.

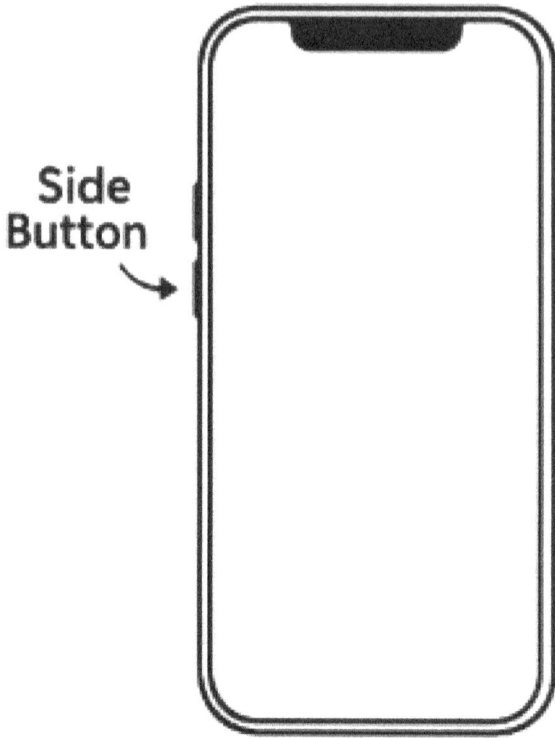

Side
Button

Press and hold the side
button until the Apple
logo appears.

Setting Up Step-by-Step

The iPhone will guide you through setup with on-screen instructions. Here's what you'll see:

1. **Swipe Up** on the "Hello" screen.

2. Choose your **Language** (for example, English).

3. Select your **Country or Region**. This ensures correct time, date, and regional settings.

4. Connect to a **Wi-Fi network**. Tap your Wi-Fi name, enter the password, and hit Join.

5. Set up **Face ID or Touch ID** (you can also skip and add later).

6. Create a **Passcode** (six digits recommended for security).

Hello

Swipe up to begin

Swipe up with your finger to start setup.

Apple ID Setup

Your Apple ID is your personal account for downloading apps, saving photos, and backing up data.

1. Enter your existing Apple ID and password.

2. If you don't have one, tap **Create Apple ID** and follow the prompts (you'll need your name, birthday, and email).

3. Agree to Apple's terms and conditions.

4. Enable **iCloud backup** when asked—this automatically saves your data.

Tip for Seniors: *If you forget passwords often, write your Apple ID and password down in a safe place.*

Sign in with
Apple ID

Apple ID

Password

Forgot password or don't have
an Apple ID?

Your Apple ID keeps all your apps, photos,
and contacts safe in iCloud.

Accessibility Setup (Highly

Recommended for Seniors)

The iPhone 17 comes with built-in features that make it

easier to use for people with vision, hearing, or motor

needs. You can enable these during setup or later in **Settings > Accessibility**.

Here are the most helpful ones:

- **Text Size & Bold Text** – Increase font size so words are easier to read.
- **VoiceOver** – Reads aloud what's on the screen.
- **Magnifier** – Turns your iPhone into a digital magnifying glass.
- **Zoom** – Lets you zoom in on any part of the screen with a simple gesture.
- **Hearing Aid Support** – Syncs with compatible devices.

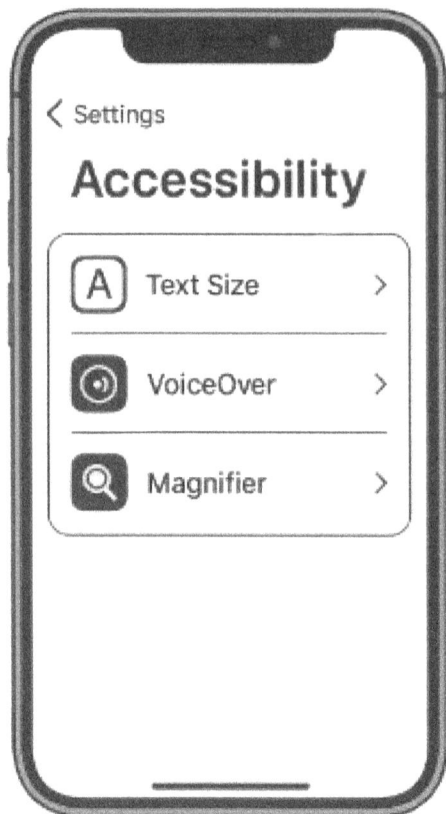

Adjust text size, use VoiceOver, or turn
on the Magnifier for easier use.

By the end of this chapter, your iPhone should be powered on, connected to Wi-Fi, secured with a passcode, and set up with your Apple ID. You'll also have accessibility tools in place so your iPhone feels comfortable from the very beginning.

Hello → | Wi-Fi
Apple ID
Passcode → Display

↑
Swipe up
to open

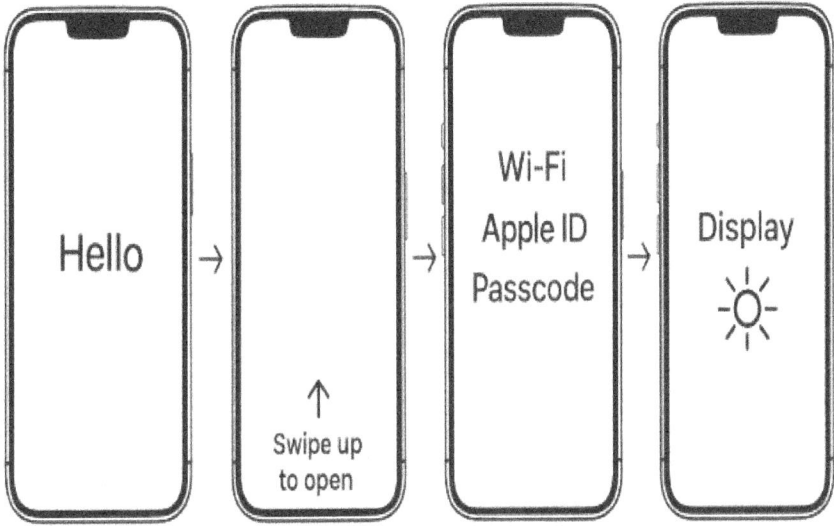

Follow the guided steps on screen. Apple makes the basics easy—you just tap, swipe, and confirm.

Chapter 3

Mastering Touch, Gestures, and Navigation

Your iPhone 17 responds to your fingers like magic. Every tap, swipe, and press is a way of telling your phone what to do. Once you learn these simple gestures, using your iPhone will feel natural and effortless.

How to Swipe, Tap, Long Press, and Use the Action Button

- **Tap** – Quickly touch your finger to the screen and lift it. Use this to open apps or select options.
- **Swipe** – Drag your finger across the screen. Swiping lets you scroll, move between screens, or reveal hidden options.

- **Long Press (Touch and Hold)** – Keep your finger on the screen for about one second. This is how you rearrange apps or open extra menus.

- **Action Button** – The new customizable button on the side of iPhone 17. You can set it to do things like:

  - Turn on the flashlight

  - Open the camera instantly

  - Start a voice memo

Tip for Seniors: Practice each gesture slowly at first. With a little repetition, your fingers will remember the motions.

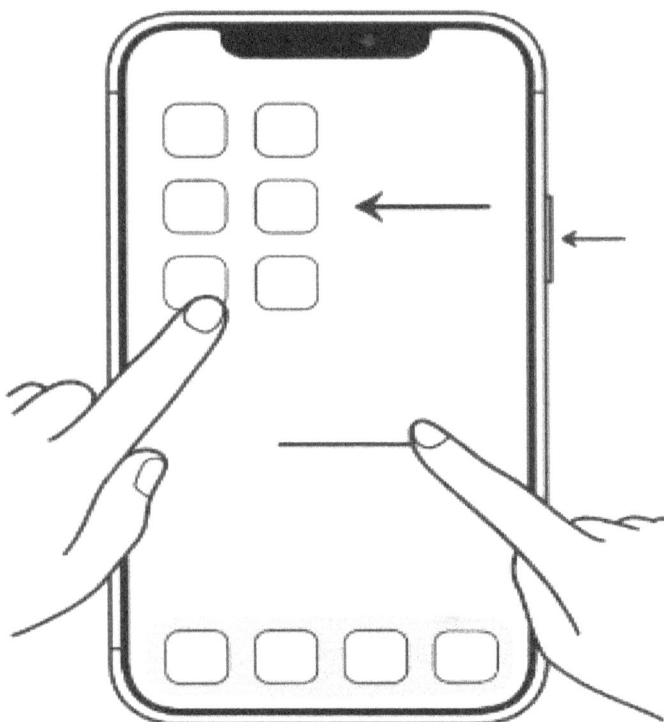

Tap to open, swipe to scroll, long press for more options. Use the Action Button for instant shortcuts.

Control Center vs Notification Center

Your iPhone has two important pull-down menus:

- **Control Center** – Swipe down from the top-right corner of the screen.

  - Here you'll find quick tools like Wi-Fi, Bluetooth, brightness, flashlight, and volume.

- **Notification Center** – Swipe down from the top-left corner or middle top.

  - Here you'll see missed calls, text messages, calendar reminders, and app alerts.

Tip: *If you swipe down and don't see what you expect, try again from the opposite corner.*

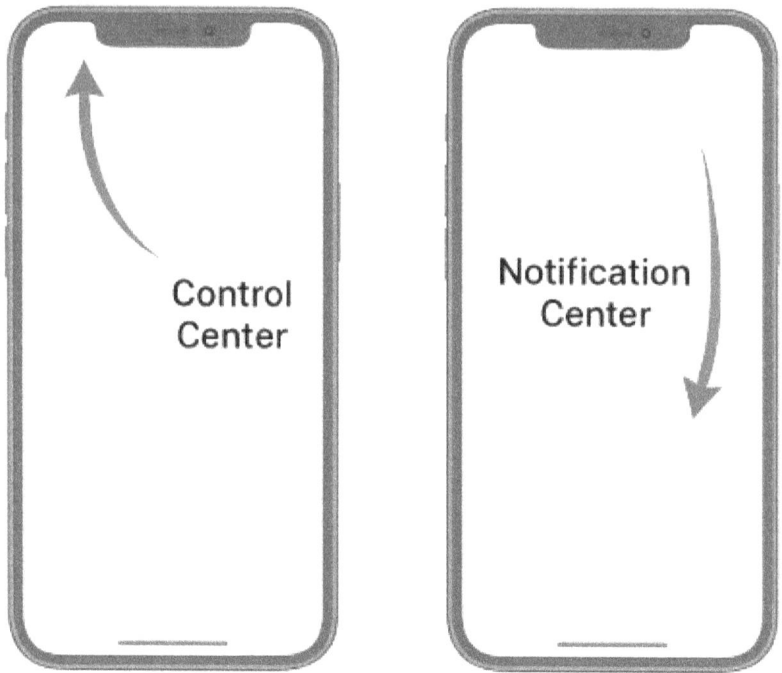

Swipe from the right corner for quick controls,
from the left for your notifications.

Multitasking & App Switching

Gone are the days of closing one app before opening
another. The iPhone 17 lets you switch between apps
instantly.

1. Swipe up from the **bottom edge of the screen**, then pause in the middle.

2. You'll see all your open apps appear as cards you can swipe through.

3. To switch, tap the one you want.

4. To close an app, swipe its card upward off the screen.

Tip for Beginners: Don't worry if you leave apps open—your iPhone is designed to manage memory efficiently.

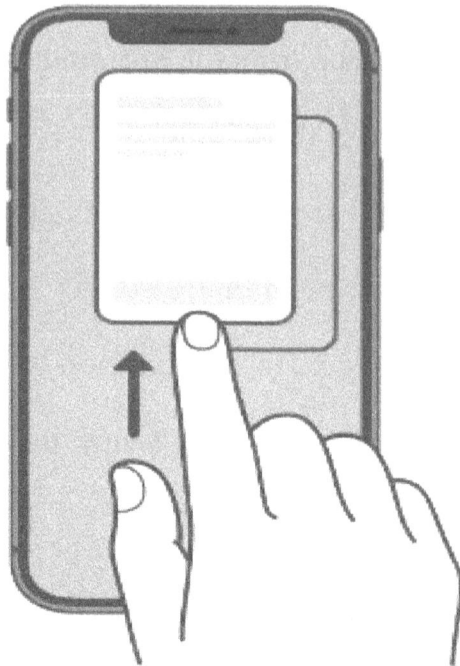

Swipe up from the bottom and
pause to see all your apps.
Swipe away to close, tap to
switch.

By the end of this chapter, you can confidently tap, swipe,

and navigate your iPhone 17. You now know how to:

- Use the Action Button for shortcuts

- Access Control Center and Notification Center

- Switch between apps like a pro

Chapter 4

Phone Calls, Messages, and Contacts Made Easy

Your iPhone 17 makes it simple to stay connected with the people who matter most. Whether you're making calls, sending messages, or saving new contacts, each step is designed to be straightforward and easy. This chapter will show you how.

Making & Receiving Calls

To Make a Call:

1. Tap the **green Phone app** on your Home Screen.
2. Tap **Keypad** at the bottom.
3. Type the number and press the green **Call** button.
4. To end a call, tap the **red End button.**

To Receive a Call:

- When your phone rings, slide the **green button** on the screen to answer.

- To decline, tap the **red button** or press the **Side Button** twice.

Tip: You can also call quickly by tapping a name in your Contacts list.

Slide to answer or tap
the red button to end
a call.

Setting Up Voicemail

1. Open the **Phone app**.

2. Tap **Voicemail** at the bottom right.

3. If it's your first time, tap **Set Up Now**.

4. Create a passcode and record your greeting (e.g., "Hi, you've reached Mary, please leave a message").

Tip for Seniors: *Keep your voicemail greeting short and clear.*

Tap Voicemail to record a personal
greeting and check messages later.

Adding Contacts

1. Open the Phone app and tap Contacts.

2. Tap the + (plus sign) in the top right.

3. Enter the person's name, phone number, and any other details (like email).

4. Tap Done to save.

Tip: *Add photos to contacts so you can recognize who's calling at a glance.*

New Contact

(+)

Name

Phone

Tap the + sign to add a new contact
with name and number.

Sending iMessages & SMS

- **iMessage (Blue bubbles):** Sent to other iPhone users over Wi-Fi or mobile data.

- **SMS (Green bubbles):** Sent to non-iPhone users through your carrier.

To Send a Message:

1. Open the **Messages app** (green speech bubble icon).

2. Tap the **compose button** in the top-right corner.

3. Type the person's name or number.

4. Tap in the message box and type your text.

5. Tap the **blue send arrow** to send.

Tip: You can also send photos, emojis, or voice messages by tapping the icons next to the text box.

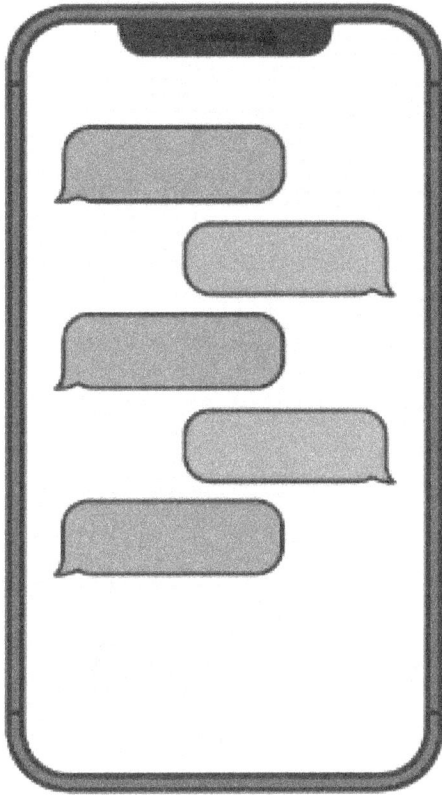

Blue = iMessage, Green = SMS.
Both keep you connected.

For Seniors: Enlarging Text & Using

Voice Dictation

- **Enlarging Text:**

1. Go to Settings > Accessibility > Display & Text Size.

2. Tap Larger Text and adjust the slider until the words are comfortable to read.

- Using Voice Dictation:

1. Open **Messages** and tap into the typing field.

2. Tap the **microphone icon** on the keyboard.

Speak your message clearly, then tap **Send**.

Tip for Seniors: Voice dictation is perfect if typing feels slow or difficult.

Tap the microphone to speak
your message instead of typing.

By the end of this chapter, you can:

- Make and answer phone calls confidently

- Set up and use voicemail

- Save and manage contacts

- Send both iMessages and SMS

- Use enlarged text and voice dictation for comfort

Chapter 5

Taking Photos & Videos Like a Pro

The iPhone 17 isn't just a phone—it's a powerful camera in your pocket. Whether you want to capture a family portrait, a beautiful night sky, or a video of your grandchild's birthday, the iPhone 17 makes it simple. Let's walk through the essentials.

iPhone 17 Camera Basics

1. Tap the **Camera app** (gray camera icon) on your Home Screen.
2. The **viewfinder** opens—what you see on screen is what your photo will look like.
3. At the bottom, swipe between **Photo, Video, Portrait, and other modes.**

4. Tap the **white circle** to take a photo.

5. Tap the **red circle** to start recording a video.

Tip for Seniors: *You don't need to rush. The iPhone camera focuses automatically, but you can also tap on the screen where you want it sharper.*

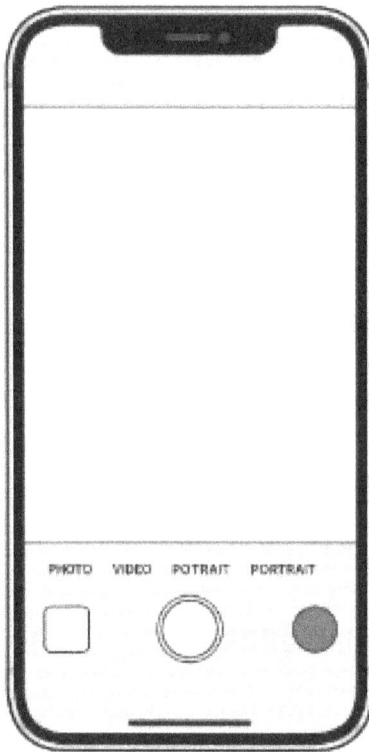

Tap the white button for photos, red for videos. Swipe left or right to switch modes.

Portrait, Night Mode, and Action Shots

- **Portrait Mode** – Creates professional photos with sharp focus on the person and a -soft blurred background. Great for family photos.
  - To use: Open Camera > Swipe to Portrait > Frame the person > Tap shutter.

- **Night Mode** – Brightens low-light photos automatically.
  - A small yellow "moon icon" shows when it's active. Just hold your phone steady while it takes the shot.

- **Action Shots** – For kids, pets, or sports.
  - Hold down the shutter button to take a burst of photos. Later, you can choose the sharpest one.

Tip: Rest your elbows on a table or hold your phone with

two hands for steadier shots.

Portrait mode makes your subject sharp and the background softly blurred.

Recording 4K Video & Cinematic Mode

- **4K Video Recording:**

1. Open Camera > Swipe to Video.

2. Look for the "HD/4K" label in the corner. Tap it to switch.

3. Tap the red button to record.

- **Cinematic Mode:**

  - Adds movie-like focus changes, smoothly switching from one subject to another.

  - To use: Swipe to Cinematic, frame your subject, and tap record.

Tip: *Cinematic mode works best with people or pets, where focus shifts between faces.*

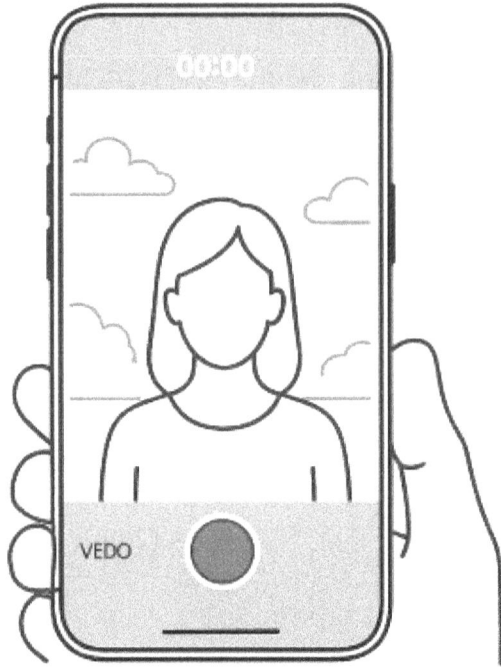

Swipe to Video, tap the red button, and capture smooth 4K memories.

Tips for Photographers:

Composition, Gridlines, and HDR

- **Composition with Gridlines** – Turn on Grid in Settings > Camera. Use the "rule of thirds" by placing your subject along the grid lines.

- **HDR (High Dynamic Range)** – Ensures bright skies and darker areas both look clear. iPhone 17 turns HDR on automatically.

- **Lighting & Angles** – Move slightly left or right to see how the light changes your photo. Sometimes small adjustments make a big difference.

Tip for Seniors: Don't be afraid to take multiple shots of the same moment. The best one is often the second or third photo.

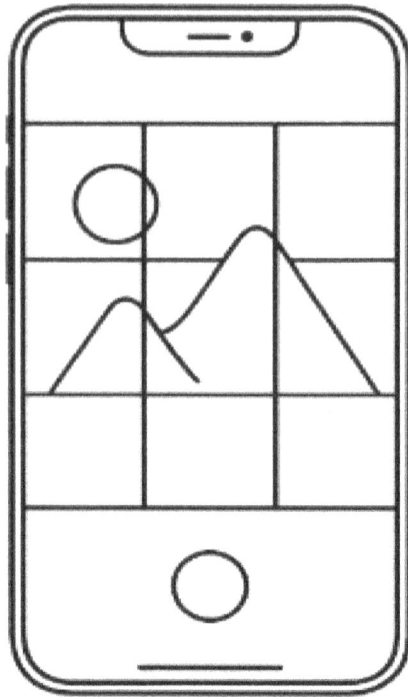

Use gridlines to balance your shot—
placing subjects off-center often
looks more natural.

By the end of this chapter, you'll know how to:

- Use the Camera app confidently

- Take stunning portraits and night photos

- Capture fast-moving action shots

- Record professional 4K and Cinematic videos

- Improve your photos with gridlines and simple composition tricks

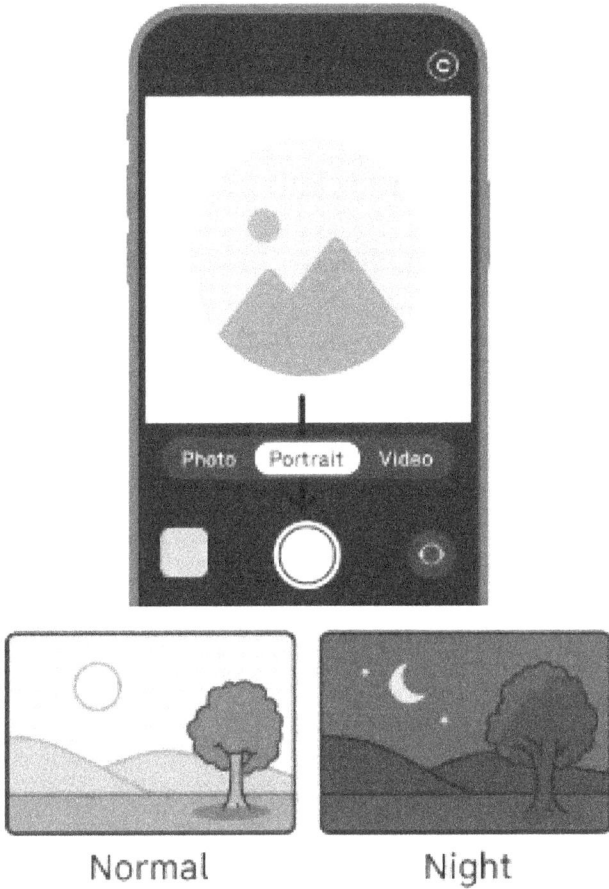

Normal   Night

Switch between Photo, Portrait, and Video by swiping along the camera bar.

Chapter 6

Internet, Apps, and the App Store

Your iPhone 17 is more than just a phone—it's your window to the world. You can browse the internet, check the news, watch videos, or download apps for almost anything you want to do. In this chapter, we'll explore how to connect online, use Safari for browsing, and safely download apps from the App Store.

Connecting to Wi-Fi & Mobile Data

To Connect to Wi-Fi:

1. Open **Settings** (gray gear icon).
2. Tap **Wi-Fi**.
3. Find your Wi-Fi network in the list and tap it.
4. Enter the password, then tap Join.

To Use Mobile Data:

1. Open **Settings > Mobile Data**.

2. Toggle the switch **ON** to allow internet access through your carrier.

Tip for Seniors: Use Wi-Fi whenever possible—it's usually faster and won't use your mobile data allowance.

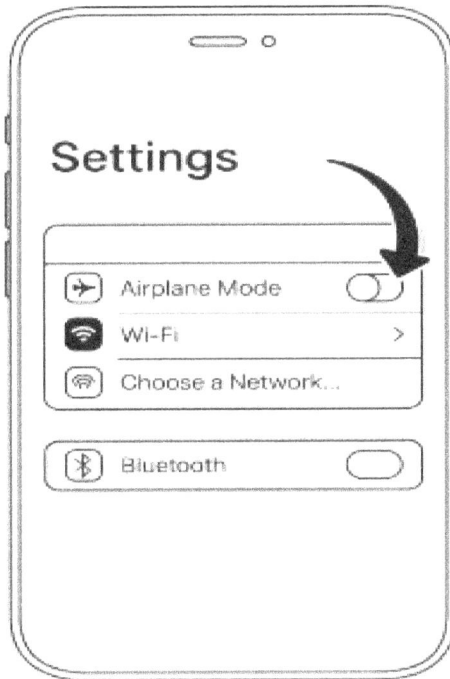

Tap Wi-Fi in Settings, choose your network, and enter the password to connect.

Browsing Safari: Tabs, Reader Mode, and Bookmarks

The Safari app (blue compass icon) is your iPhone's built-in internet browser.

- **Opening a Website:** Tap Safari, type a website (e.g., www.mayobook.com) in the address bar, and tap Go.

- **Using Tabs:** Tap the square icon in the bottom corner to switch between multiple open web pages.

- **Reader Mode:** If a web page feels cluttered, tap the "AA" button in the address bar and select Show Reader View. This shows only the main text and images, making it easier to read.

- **Bookmarks:** Tap the share icon (square with arrow) and select Add Bookmark to save a favorite site.

Tip for Beginners: Reader Mode is especially useful for seniors because it removes ads and enlarges text.

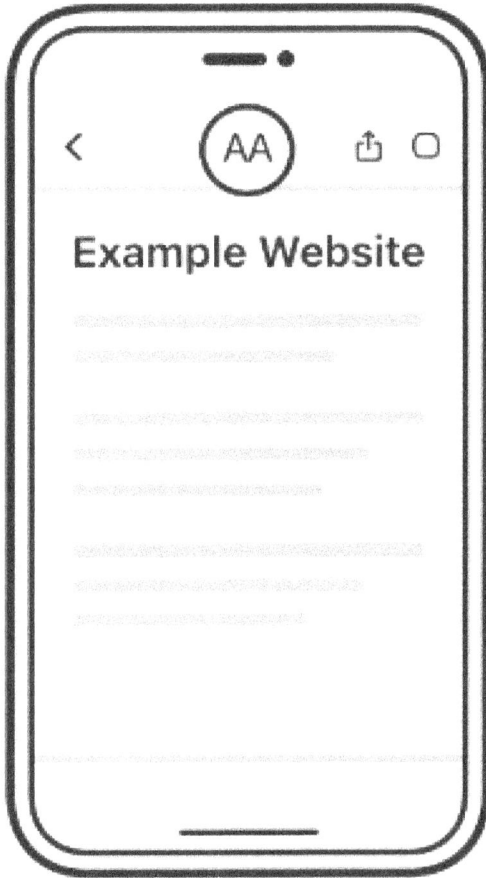

Tap AA in the address bar to switch to Reader View for a cleaner page.

Downloading Apps Safely from the App Store

The App Store (blue "A" icon) is where you can download apps like WhatsApp, YouTube, Kindle, or games.

1. Tap the **App Store** icon on your Home Screen.
2. Tap the **Search** tab (magnifying glass icon at the bottom).
3. Type the app name (e.g., "Zoom") and press Search.
4. Tap **Get** (or the price if it's paid).
5. Confirm with **Face ID, Touch ID, or your passcode**.

Safety Tip: Always download apps only from the App Store. Avoid clicking on random pop-ups or outside links.

Tap GET to download an app.
You may need Face ID, or a
passcode to confirm.

By the end of this chapter, you can:

- Connect to Wi-Fi and mobile data

- Browse the internet with Safari confidently

- Use Reader Mode and bookmarks for easier reading

- Download safe and useful apps from the App Store

Chapter 7

Staying Connected (Email, Social, and FaceTime)

Your iPhone 17 helps you stay close to family and friends, no matter where they are in the world. With email, video calls, and popular apps like WhatsApp or Zoom, you can share moments, have face-to-face conversations, and stay updated on social media with just a few taps.

Setting Up Gmail, Yahoo, or iCloud Mail

To Add an Email Account:

1. Open **Settings** (gray gear icon).

2. Scroll down and tap **Mail**.

3. Select **Accounts > Add Account**.

4. Choose your email provider (Gmail, Yahoo, Outlook, or iCloud).

5. Enter your email address and password.

6. Tap **Next** and wait for your account to sync.

Tip for Seniors: *If typing passwords feels difficult, use Voice Dictation by tapping the microphone on the keyboard.*

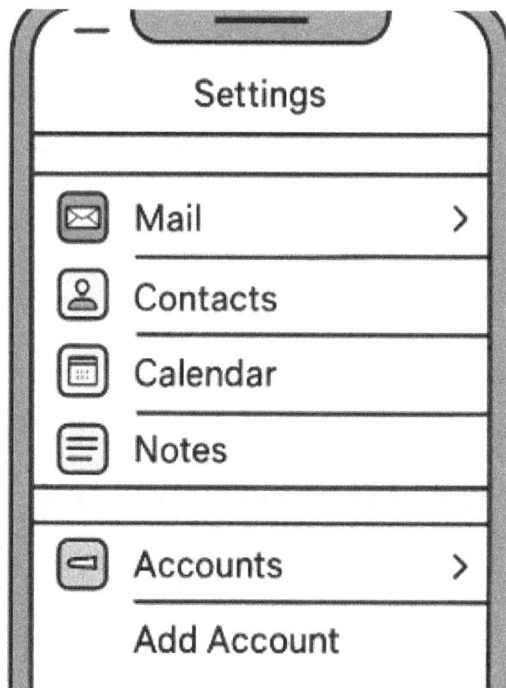

Add your Gmail, Yahoo, or iCloud account in Settings > Mail > Accounts.

FaceTime Calls (Video + Audio)

FaceTime is Apple's built-in video and audio calling app. It works over Wi-Fi or mobile data and is free between Apple devices.

To Make a FaceTime Call:

1. Open the **FaceTime app** (green camera icon).
2. Tap the + **(plus)** button.
3. Type the name, email, or phone number of the person.
4. Tap **Video** for a video call or **Audio** for voice only.

To Answer a FaceTime Call:

- Tap the **green button** to accept, or the **red button** to decline.

Tip: Hold your phone in front of your face so the camera captures you clearly. Good lighting makes video calls more enjoyable.

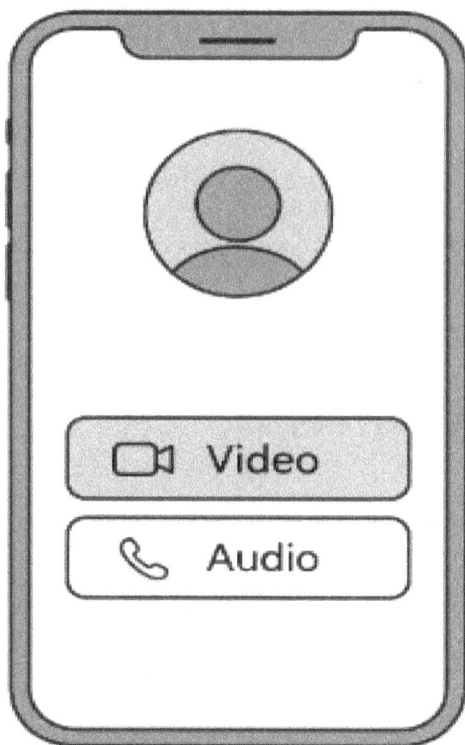

Tap Video for a face-to-face call
or Audio for a voice call.

WhatsApp, Zoom, and Social Media

Basics

Not everyone uses FaceTime, but apps **like WhatsApp,**

Zoom, Facebook, and Instagram make it easy to connect

with everyone, even non-iPhone users.

- **WhatsApp:** Free messaging and video calls worldwide. Great for family chats.
- **Zoom:** Video meetings—popular for church groups, classes, and family gatherings.
- **Facebook & Instagram:** Stay updated with photos, messages, and news from family and friends.

To Get These Apps:

1. Open the **App Store**.
2. Tap **Search** and type the app's name.
3. Tap **Get** (or the download icon).
4. Open the app and follow the setup instructions.

Tip for Seniors: *Don't be afraid of social media—it's just another way to keep in touch. Start small by following close friends and family.*

Download apps like WhatsApp, Zoom, or Facebook from the App Store to stay connected with everyone.

By the end of this chapter, you'll know how to:

- Add and use an email account on your iPhone

- Make and receive FaceTime video or audio calls

- Download and use popular apps like WhatsApp, Zoom, and Facebook

Tap the green camera to start Face-Time. Add multiple people for group calls.

Chapter 8

Personalization & Accessibility

Your iPhone 17 doesn't have to look or sound like everyone else's. You can change wallpapers, ringtones, and even rearrange your Home Screen so it feels like your phone. Apple also includes powerful accessibility features to make the iPhone easier to see, hear, and use for seniors or anyone with special needs.

Changing Wallpapers and Ringtones

To Change Your Wallpaper:

1. Open **Settings > Wallpaper**.
2. Tap **Choose a New Wallpaper**.

3. Pick from Apple's built-in images, dynamic wallpapers, or your own photos.

4. Tap **Set** and choose whether it applies to the Lock Screen, Home Screen, or both.

To Change Your Ringtone:

1. Go to **Settings > Sounds & Haptics**.

2. Tap **Ringtone**.

3. Choose from the default tones or tap **Tone Store** to download new ones.

Tip for Seniors: Choose bright, high-contrast wallpapers and louder ringtones so they're easy to see and hear.

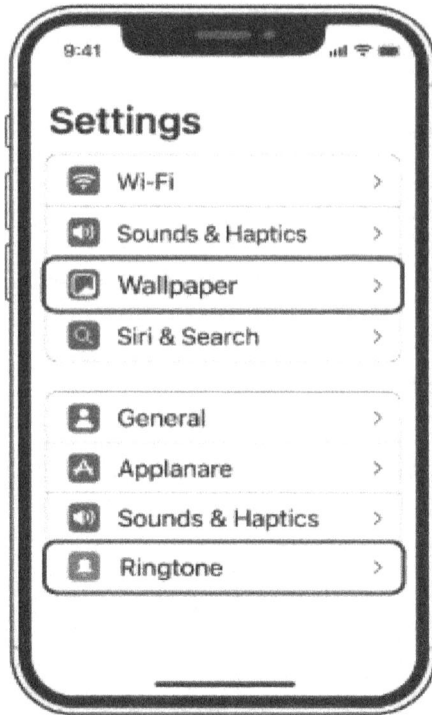

Make your iPhone yours—choose
wallpapers and ringtones that
suit your style.

Customizing Home Screen Widgets

Widgets give you quick information at a glance—like weather, calendar events, or battery life.

To Add a Widget:

1. Touch and hold an empty space on the Home Screen until apps jiggle.

2. Tap the + **icon** in the top left.

3. Scroll to find a widget (Weather, Calendar, News, etc.).

4. Tap **Add Widget**.

5. Drag it into place and tap **Done**.

Tip: *Place the Clock, Calendar, or Weather widget at the top of your Home Screen for quick daily updates.*

Widgets show useful info at glance—
add them by tapping the + icon
on your Home Screen.

Accessibility Options (VoiceOver, Hearing Aids, Magnifier, Zoom)

The iPhone 17 includes tools that make it easier to see, hear, and interact with your device.

- **VoiceOver:** Reads aloud everything on the screen. Great for users with vision challenges.

  - Enable via Settings > Accessibility > VoiceOver.

- **Hearing Aid Compatibility:** Pair supported hearing aids for clearer calls.

- **Magnifier:** Turns your iPhone into a magnifying glass—perfect for reading small text.

  - Triple-press the Side Button to activate.

- **Zoom:** Lets you zoom in on any part of the screen by double-tapping with three fingers.

Tip for Seniors: Turn on Bold Text and Larger Text under Accessibility to make everything easier to read.

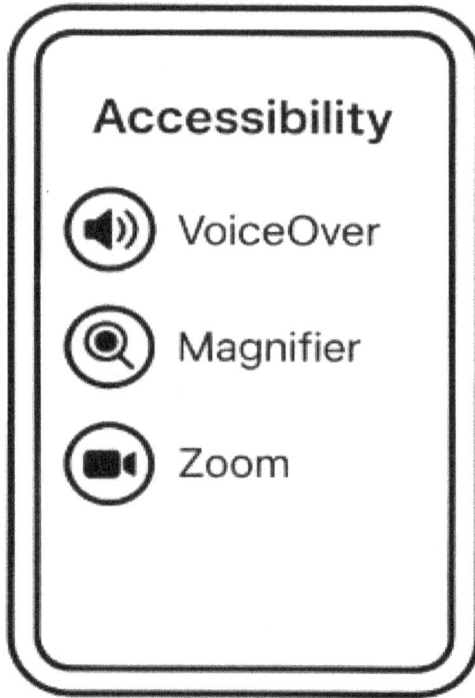

Accessibility

🔊 VoiceOver

🔍 Magnifier

📹 Zoom

Adjust text size, use Magnifier,
or turn on VoiceOver for
easier reading and navigation.

By the end of this chapter, you'll know how to:

- Change wallpapers and ringtones

- Add widgets to your Home Screen

- Use accessibility tools like VoiceOver, Magnifier, and Zoom

Make your iPhone yours—choose from Apple wallpapers or use your own photos.

Chapter 9

Security & Privacy Essentials

Your iPhone 17 is designed with strong security features to protect your information. Whether it's your photos, bank details, or simple text messages, Apple gives you tools to keep everything safe from prying eyes and online scams. In this chapter, we'll explore Face ID, passcodes, two-factor authentication, app permissions, and scam protection tips—especially important for seniors.

Face ID & Passcodes

Setting Up Face ID:

1. Open Settings > Face ID & Passcode.
2. Enter your existing passcode.
3. Tap Set Up Face ID.

4. Hold your iPhone at eye level and move your head slowly in a circle to scan your face.

Tip: *Glasses or masks? You can set up an **Alternate Appearance** to make unlocking smoother.*

Creating a Strong Passcode:

1. Go to **Settings > Face ID & Passcode > Change Passcode.**

2. Choose a **6-digit numeric code** or tap **Passcode Options** for a longer, stronger one.

3. Avoid using easy codes like 123456 or your birthday.

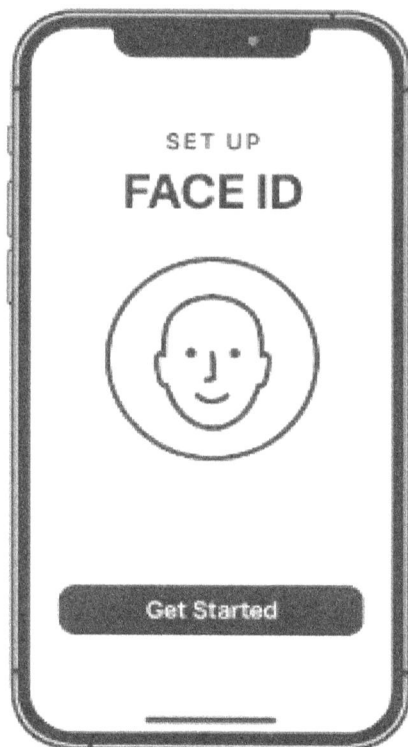

Use Face ID and a strong passcode to keep your iPhone secure.

Two-Factor Authentication

Two-factor authentication (2FA) adds an extra layer of
security to your Apple ID. Even if someone guesses your
password, they can't log in without a code sent to your

iPhone.

To Turn It On:

1. Go to **Settings > [Your Name] > Password & Security**.

2. Tap **Turn On Two-Factor Authentication**.

3. Confirm your trusted phone number.

4. Each time you sign in on a new device, you'll receive a 6-digit code.

Tip for Seniors: Think of it as a lock and key—your password is the lock, the 6-digit code is the key.

Enter Verification
Code

Two-factor authentication
sends a 6-digit code to
your iPhone whenever you
log in.

Location Services and App Permissions

Many apps request permission to use your location or access your photos, microphone, or camera. You should always review these requests carefully.

To Manage Location Settings:

1. Open **Settings > Privacy & Security > Location Services**.

2. Toggle apps individually: choose **Never, Ask Next Time, or While Using the App**.

To Review App Permissions:

1. Go to **Settings > Privacy & Security**.

2. Tap categories like **Camera, Microphone, or Photos**.

3. Choose which apps can access each feature.

Tip: *If an app asks for permissions it doesn't need (like a calculator requesting camera access), deny it.*

Location Services

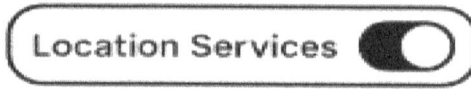

Location Services 🔘

< LOCATION

Never

Ask Next Time
Or When I Share

While Using
the App   ✓

Allow apps to use your location
only when necessary—choose
'While Using the App" for safety.

Scam Protection for Seniors

Scammers often target seniors with phone calls, texts, or

fake pop-ups. Protect yourself with these iPhone tools:

- **Silence Unknown Callers:** Go to **Settings > Phone > Silence Unknown Callers**. Unknown numbers go straight to voicemail.

- **Message Filtering:** Enable Filter Unknown Senders in Settings > Messages.

- **Fraudulent Websites Warning:** In Settings > Safari, keep Fraudulent Website Warning turned on.

- **Never Tap Unknown Links:** If you don't recognize a text or email, delete it.

Tip for Seniors: Remember—banks, Apple, or government agencies will never ask you for your password or payment info over text or email.

Block unknown callers and ignore suspicious links to stay safe from scams.

By the end of this chapter, you can:

- Set up Face ID and a strong passcode

- Protect your Apple ID with two-factor authentication

- Control app permissions and location settings

- Recognize and block scam attempts

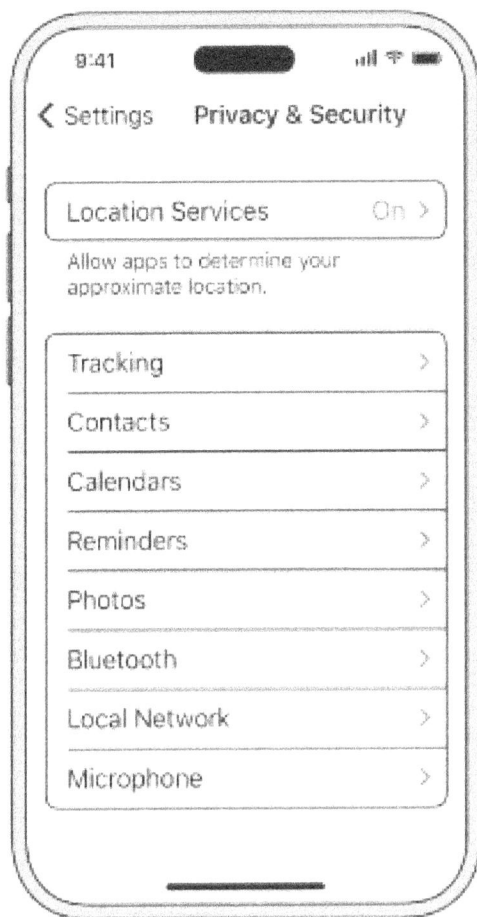

9:41

< Settings   Privacy & Security

Location Services   On >

Allow apps to determine your
approximate location.

Tracking   >

Contacts   >

Calendars   >

Reminders   >

Photos   >

Bluetooth   >

Local Network   >

Microphone   >

Only allow location access when needed—
choose 'While Using App.'

Chapter 10

Switching from Android to iPhone 17

Moving from Android to iPhone may feel like a big step, but Apple makes the process smoother than ever. With the **Move to iOS app**, you can transfer your contacts, photos, messages, and more in just a few taps. In this chapter, we'll guide you through the process and point out the main differences between Android and iOS so you can adjust quickly.

Using Move to iOS App

Apple's **Move to iOS app** transfers your most important data wirelessly from your old Android device to your new iPhone 17.

Step-by-Step:

1. On your **Android device**, open the **Google Play** Store and download the app **Move to iOS**.

2. On your **iPhone 17**, begin the setup process.

3. When you reach the **Apps & Data** screen, select **Move Data from Android**.

4. **A 6-digit or 10-digit code** will appear on your iPhone.

5. On your Android phone, open Move to iOS, enter the code, and connect both devices over Wi-Fi.

6. Select what you want to transfer (contacts, photos, messages, apps).

7. Wait for the process to finish, then continue setting up your iPhone.

Tip for Seniors: Keep both phones plugged into power during the transfer. Large photo collections may take a while.

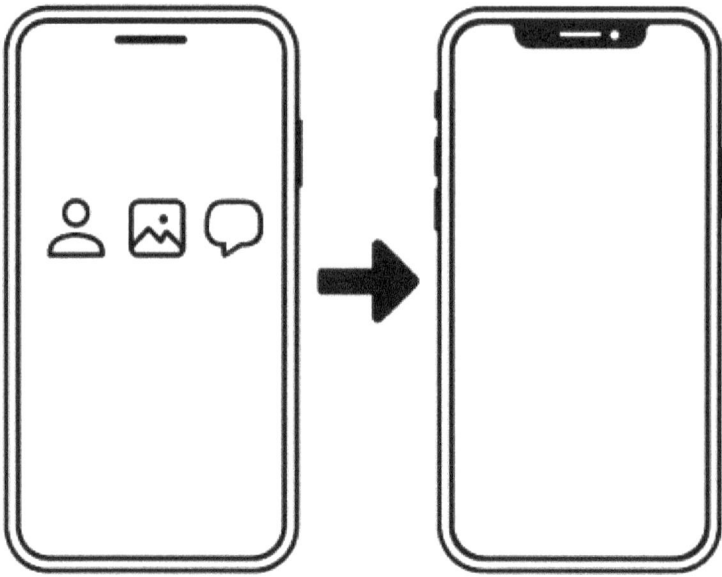

Use Move to iOS to copy contacts, photos, and messages from Android to iPhone.

Migrating Contacts, Photos, and Messages

If you didn't use Move to iOS, you can still transfer items individually:

- **Contacts:** Save your contacts to your Google account on Android. Then, on iPhone, go to Settings > Mail > Accounts > Add Account > Google. Turn on Contacts sync.

- **Photos:** Install the Google Photos app on iPhone. Log in with your Google account and download your pictures.

- **Messages & WhatsApp Chats:** Many messaging apps (like WhatsApp) now support chat history transfer between Android and iOS. During WhatsApp setup on iPhone, follow the prompts to import chats.

Tip: Even after moving, keep your Android handy for a few days to double-check that nothing was missed.

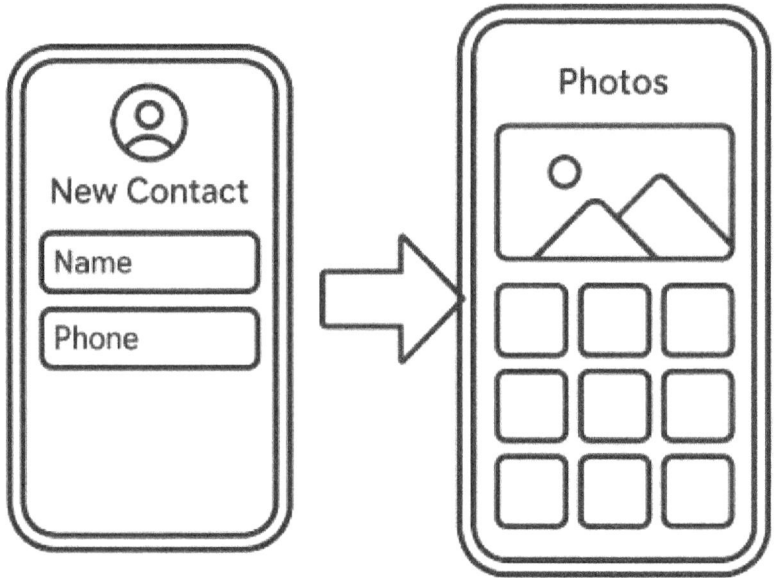

You can also sync contacts through
Google and move photos with
Google Photos.

Adjusting to iOS after Android (Key

Differences Simplified)

Switching systems means some things will look different.

Here are the biggest differences explained simply:

- **Back Button:** Android often has a dedicated back button. On iPhone, just swipe from the left edge of the screen or tap the "< Back" label at the top left.

- **App Drawer vs Home Screen:** Android separates the app drawer from the Home Screen. On iPhone, every app appears on the Home Screen, but you can organize them into folders or use the App Library.

- **Customization:** iPhones use Widgets and Wallpapers for personalization, but keep menus simpler than Android.

- **Charging Cable:** The iPhone 17 uses USB-C, which may already be familiar from newer Android phones.

- **System Updates:** iPhones receive iOS updates for years, directly from Apple, so your phone stays secure longer.

Tip for Seniors: Don't be discouraged by the differences. Within a week of daily use, most Android users find iOS simpler and more predictable.

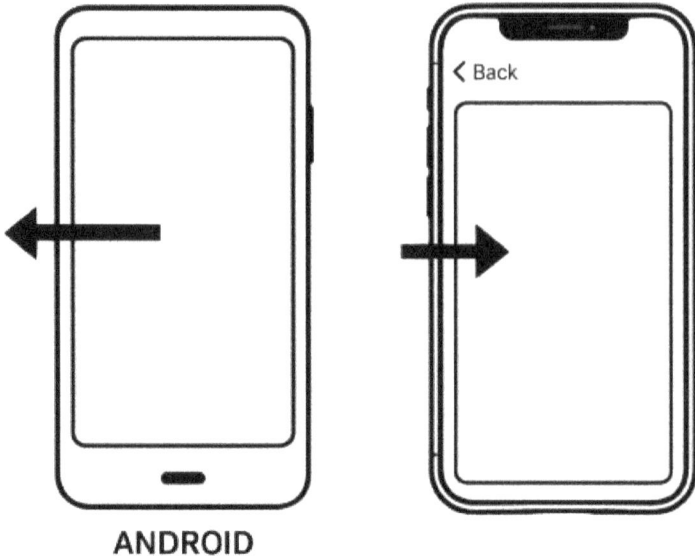

ANDROID

On iPhone, swipe from the left edge or tap Back
at the top instead of using a button.

By the end of this chapter, you'll know how to:

- Transfer your data with Move to iOS

- Bring over contacts, photos, and messages

- Adjust quickly to the differences between Android
  and iOS

With Move to iOS, your data
transfers securely in minutes.

Chapter 11

Troubleshooting Made Simple

Even the smartest phone has its hiccups. Don't worry—most problems with the iPhone 17 have quick, simple fixes you can do yourself. In this chapter, we'll look at common issues like Wi-Fi drops, Bluetooth glitches, frozen screens, and how to restart or reset settings without losing your data.

Common Problems & Fixes

Wi-Fi Won't Connect

1. Go to **Settings** > **Wi-Fi.**
2. Toggle Wi-Fi **OFF**, then back **ON.**
3. If still stuck, tap your Wi-Fi name, select **Forget This Network**, then reconnect.

Tip: *Restarting your router at home can also solve most*

Wi-Fi problems.

Bluetooth Not Pairing

1. Go to **Settings > Bluetooth**.

2. Find your device in the list. Tap the i icon and choose **Forget This Device**.

3. Try pairing again by turning Bluetooth off, then on.

Tip: Keep devices close together during pairing—within a few feet.

Frozen or Unresponsive Screen

- **Quick Fix:** Press and release **Volume Up**, then **Volume Down**, then **hold the Side Button** until the Apple logo appears.

- This is called a **force restart** and won't erase your data.

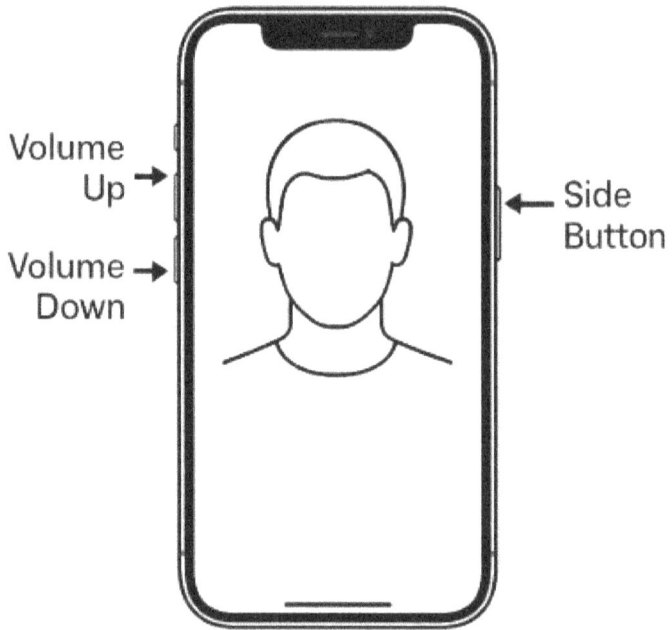

Press Volume Up, then Volume Down, then hold the Side Button until the Apple logo appears.

How to Restart or Force Restart

Normal Restart:

1. Press and hold the **Side Button + Volume Up** until you see **Slide to Power Off**.

2. Drag the slider.

3. To turn back on, hold the Side Button until the Apple logo appears.

Force Restart (when phone is frozen):

1. Press **Volume Up** once.

2. Press **Volume Down** once.

3. Hold the **Side Button** until the Apple logo appears.

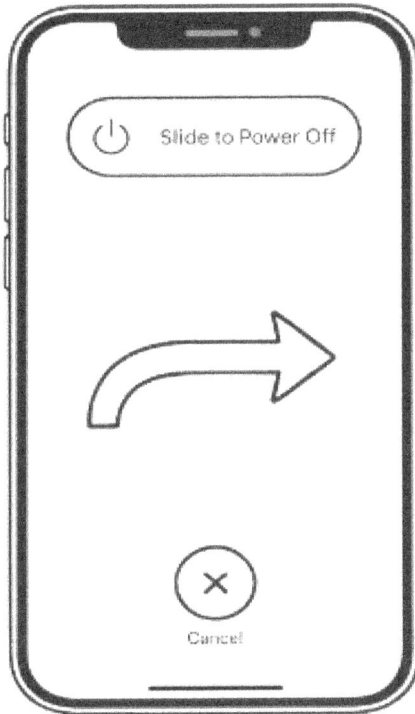

Slide to power off. If frozen, use the button sequence for a force restart.

Resetting Settings Without Erasing Data

Sometimes, a misbehaving iPhone just needs a "fresh start" for its settings. This resets preferences but does not erase photos, apps, or contacts.

To Reset Settings:

1. Go to **Settings > General > Transfer or Reset iPhone.**
2. Tap **Reset > Reset All Settings**.
3. Enter your passcode to confirm.

Tip for Seniors: This is safe—you won't lose your personal data, but Wi-Fi passwords and wallpaper settings will return to default.

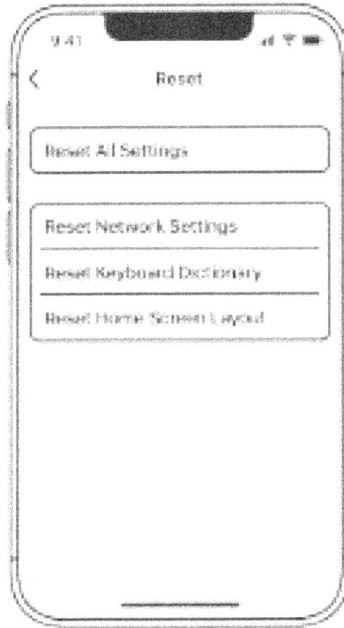

Reset All Settings fixes many problems
without erasing your data.

By the end of this chapter, you'll know how to:

- Solve Wi-Fi and Bluetooth problems

- Restart and force restart your iPhone 17

- Reset settings safely without losing data

Chapter 12

Hidden Features & Tips You'll Love

Your iPhone 17 has clever little secrets that can make everyday tasks faster and easier. These hidden features aren't always obvious, but once you learn them, you'll wonder how you ever lived without them.

Back Tap Shortcuts

Did you know you can tap the back of your iPhone to trigger shortcuts? This hidden trick is called Back Tap.

To Turn It On:

1. Go to **Settings > Accessibility > Touch**.
2. Scroll down and tap **Back Tap**.
3. Choose **Double Tap** or **Triple Tap**.

4. Assign an action like **Flashlight, Screenshot, or Volume Up**.

Tip: *For seniors, setting Back Tap to turn on the flashlight makes nighttime use much easier.*

Double or triple tap the back of your iPhone to launch shortcuts like

Action Button Customization

The iPhone 17 introduces the Action Button—a customizable side button that replaces the old mute switch.

To Customize:

1. Open **Settings** > **Action Button**.
2. Select what you'd like it to do:
   - Camera
   - Voice Memo
   - Flashlight
   - Magnifier
   - Shortcut (like opening your favorite app)

Tip for Seniors: Assign the Action Button to Camera for quick photo-taking without unlocking menus.

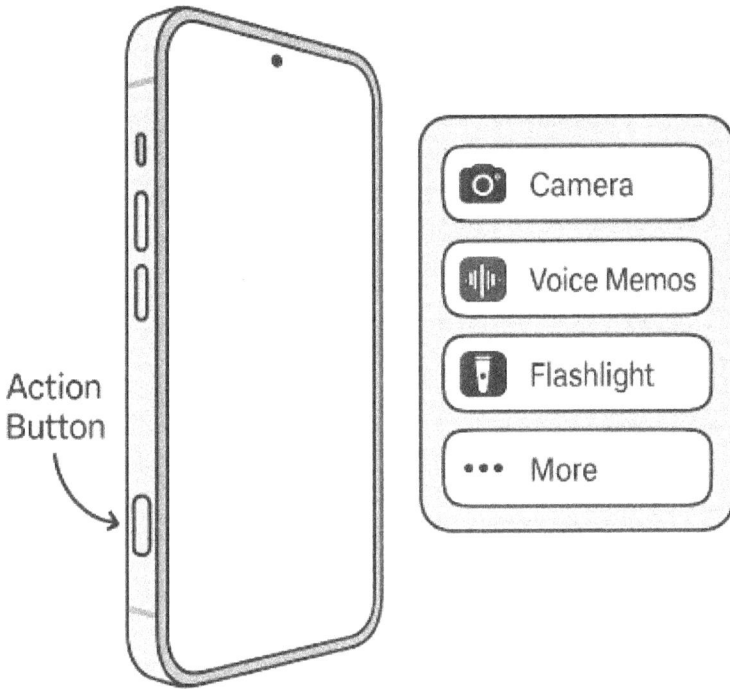

Action
Button

Camera

Voice Memos

Flashlight

... More

Set the Action Button to instantly open
Camera, Voice Memos, Flashlight, or more.

Live Text & Translate in Photos

Your iPhone can now read text directly from photos—
perfect for saving phone numbers, recipes, or documents.

Using Live Text:

1. Open the **Camera app** and point it at text (like a sign or recipe).

2. A **yellow frame** will appear around the text.

3. Tap **Copy, Translate, or Share** as needed.

Using Translate:

- Take a photo of foreign text, then tap **Translate**. Your iPhone will show the English version instantly.

Tip: Seniors traveling abroad can use this to read menus, street signs, or medication labels.

Live Text lets you copy or translate words directly from photos.

Siri Commands for Everyday Life

Siri, your iPhone's voice assistant, can save time with simple voice commands.

Everyday Siri Commands:

- "Call John."

113

- "Send a message to Sarah: I'll be there soon."

- "What's the weather tomorrow?"

- "Set an alarm for 7 AM."

- "Remind me to take my medicine at 8 PM."

Tip for Seniors: *Siri is like having a personal assistant. Speak naturally—no need to memorize commands.*

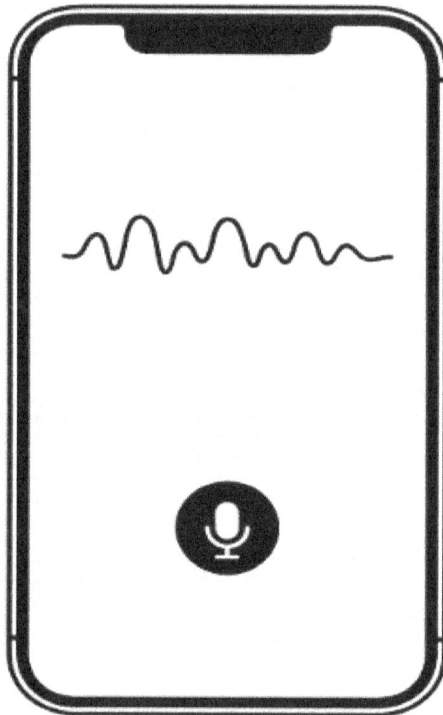

Ask Siri to call, text, set reminders, or check the weather—just speak naturally.

By the end of this chapter, you'll know how to:

- Use Back Tap for instant shortcuts

- Customize the new Action Button

- Extract and translate text with Live Text

- Let Siri handle reminders, calls, and everyday questions

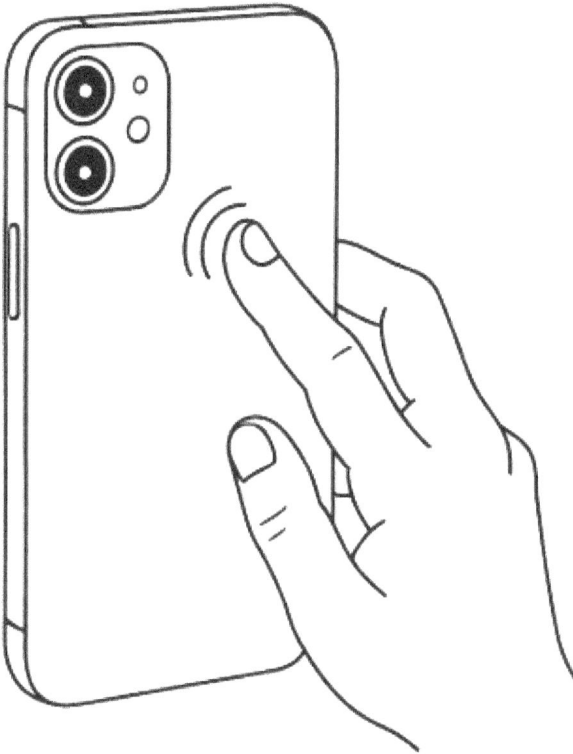

Assign Back Tap to flashlight or screenshot for one-tap convenience

Chapter 13

Must-Have Apps for Seniors & Beginners

The iPhone 17 becomes even more powerful when you add the right apps. For seniors and beginners, the best apps are those that improve health, safety, learning, and daily living. Let's explore some must-have apps that are easy to download, safe to use, and designed to make life simpler.

Health & Safety Apps

Medications & Reminders

- MediSafe – Helps you remember when to take your medicine with clear notifications.
- Apple Health – Already built-in, tracks your steps, heart rate, and activity.

Fall Detection & SOS

- Apple Health Emergency SOS – On iPhone and Apple Watch, automatically alerts emergency services if you fall or press the SOS button.

Emergency Contact Apps

- ICE (In Case of Emergency) – Stores medical info and emergency contacts for quick access by doctors or first responders.

Tip for Seniors: Always set up **Medical ID** in the Health app. It can display your allergies, medications, and emergency contacts right on the Lock Screen.

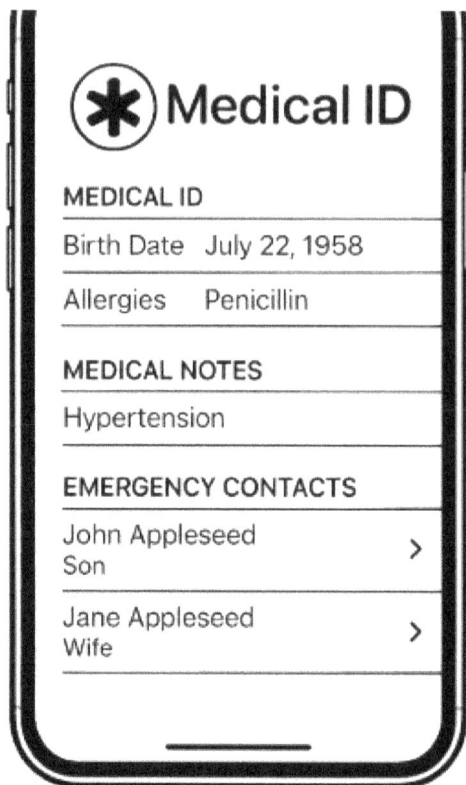

Set up Medical ID in the Health app

Learning & Reading Apps

Kindle (Amazon)

- Access thousands of eBooks in large-print mode.
  Great for seniors who prefer adjustable text size.

Audible

- Listen to audiobooks—perfect for bedtime stories, traveling, or those with vision difficulties.

Duolingo

- Learn a new language with fun, short lessons. Many seniors use it to keep their minds sharp.

Tip: These apps allow you to adjust font size or play speed, making learning comfortable at your pace.

Lifestyle Apps

Banking & Bills

- **Chase Mobile, Bank of America, PayPal** (examples)—let you check balances, pay bills, and transfer money securely.

Shopping

- **Amazon** – Order essentials with one tap.

- **Instacart** – Get groceries delivered right to your door.

Social Media

- **Facebook** – Stay connected with family updates and photos.
- **Instagram** – See your grandchildren's pictures and short videos.

Tip for Seniors: *Be careful about scams on social media. Only accept friend requests from people you know personally.*

Download lifestyle apps like
Facebook, Kindle, and Amazon to
stay connected and shop easily.

- By the end of this chapter, you'll know how to:

- Install health and safety apps to protect yourself

- Use reading and learning apps for fun and mental sharpness

- Manage money, shopping, and social connections with lifestyle apps

Bonus Section A – iOS 18 Tips & Tricks

iOS 18 introduces small but powerful upgrades that make the iPhone 17 even more useful.

New Widgets & Lock Screen Customizations

- Add **interactive widgets** to control music, lights, or reminders right from your Home Screen.
- Lock Screen now supports **bigger fonts, new colors, and more wallpaper choices**—making it easier for seniors to read.

Tip: Place your Calendar and Weather widgets on the Lock Screen so you see your day at a glance.

Customize your Lock Screen with bigger fonts and interactive widgets.

Smarter Siri

- Siri now understands **follow-up questions** without repeating "Hey Siri" every time.

- Example: "What's the weather tomorrow?" → "And what about Friday?"

- Works better offline for basic tasks like setting alarms or opening apps.

Tip for Seniors: *Try asking Siri to "Read my new messages" or "Turn on the flashlight."*

Siri now understands follow-up questions and works even offline.

Improved Battery Management

- iOS 18 learns your daily charging routine and reduces wear on the battery.

- **Optimized Charging**: Keeps your battery at 80% until just before you unplug.

- Battery Health section shows detailed information.

Tip: Turn on Low Power Mode (Settings > Battery) if you're out all day.

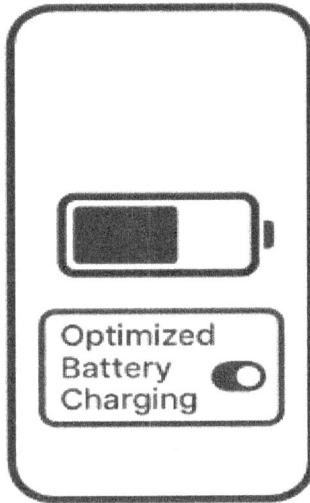

Optimized Charging helps
your battery last longer.

Bonus Section B – Senior Quick Reference Guide

This one-page **illustrated cheat sheet** is designed as a memory aid.

Essential Icons:

- **Phone** – Calls
- **Messages** – Texts
- **Safari** – Internet
- **Music** – Songs
- **Camera** – Photos/Videos
- **Settings** – Adjust iPhone

Essential Gestures:

- **Tap** – Open
- **Swipe** – Move/Scroll
- **Pinch Out/In** – Zoom
- **Swipe from Left Edge** – Go back

Tip: *Print this page or keep it bookmarked in your Kindle version for quick access.*

Quick-reference guide: know your essential icons and gestures at a glance

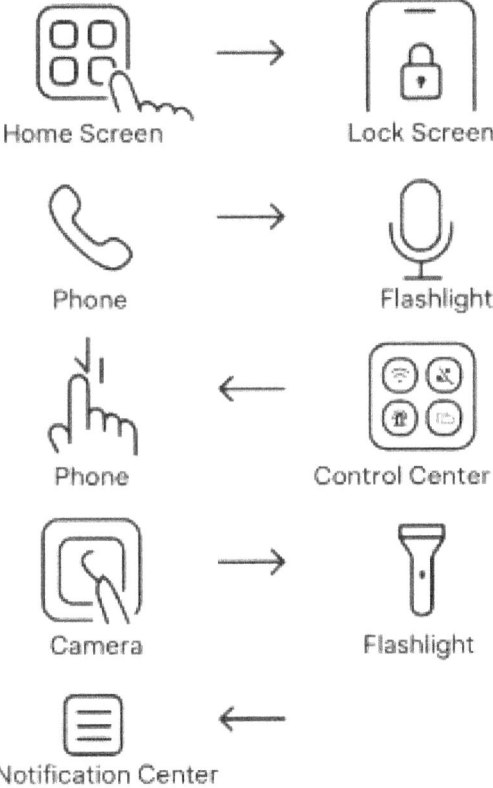

Home Screen → Lock Screen

Phone → Flashlight

Phone ← Control Center

Camera → Flashlight

Notification Center ←

Bonus Section C – Glossary of iPhone Terms

A simple glossary makes technology less intimidating.

- **AirDrop** – Share files instantly with nearby Apple devices.

- **Apple ID** – Your personal account for apps, email, and iCloud.

- **iCloud** – Online storage that saves your photos, contacts, and documents.

- **App Store** – Where you download apps (like WhatsApp or Kindle).

- **Control Center** – Swipe down from the top-right to quickly adjust Wi-Fi, brightness, or volume.

- **Widgets** – Small boxes on your screen that show information (like weather or calendar).

- **Siri** – Your voice assistant that responds to commands like "Call Sarah."

- **Face ID** – Unlocks your iPhone using your face.

- **Two-Factor Authentication** – An extra security step that sends a code to your phone.

Tip for Seniors: *Refer back to this glossary whenever a new word confuses you—over time, these terms will feel natural.*

With these bonus sections, your guide isn't just a manual—it becomes a complete toolkit for seniors and beginners. You've given them:

- iOS 18 updates for the newest features

- A quick one-page reference for daily use

- A glossary to remove tech confusion

Acknowledgement

To every aspiring photographer and filmmaker who dares to pick up a camera and tell a story, this book is for you. Special thanks to my family and friends for their encouragement, and to the creative community whose passion inspires me daily. Your support made this guide possible.

www.ingramcontent.com/pod-product-compliance
Lightning Source LLC
Chambersburg PA
CBHW031900200326
41597CB00012B/488